AF798077

Section 1

Conversion Factors and Mathematical Symbols*

James O. Maloney, Ph.D., P.E. *Emeritus Professor of Chemical Engineering, University of Kansas; Fellow, American Institute of Chemical Engineering; Fellow, American Association for the Advancement of Science; Member, American Chemical Society; Member, American Society for Engineering Education*

CONVERSION FACTORS

CONVERSION OF VALUES FROM U.S. CUSTOMARY UNITS TO SI UNITS

*Much of the material was taken from Sec. 1. of the fifth edition. The contribution of Cecil H. Chilton in developing that material is acknowledged.

TABLE 1-1 SI Base and Supplementary Quantities and Units

Quantity or "dimension"	SI unit	SI unit symbol ("abbreviation"); Use roman (upright) type
Base quantity or "dimension"		
length	meter	m
mass	kilogram	kg
time	second	s
electric current	ampere	A
thermodynamic temperature	kelvin	K
amount of substance	mole*	mol
luminous intensity	candela	cd
Supplementary quantity or "dimension"		
plane angle	radian	rad
solid angle	steradian	sr

*When the mole is used, the elementary entities must be specified; they may be atoms, molecules, ions, electrons, other particles, or specified groups of such particles.

TABLE 1-2*a* Derived Units of SI that Have Special Names

Quantity	Unit	Symbol	Formula
frequency (of a periodic phenomenon)	hertz	Hz	l/s
force	newton	N	(kg·m)/s^2
pressure, stress	pascal	Pa	N/m^2
energy, work, quantity of heat	joule	J	N·m
power, radiant flux	watt	W	J/s
quantity of electricity, electric charge	coulomb	C	A·s
electric potential, potential difference, electromotive force	volt	V	W/A
capacitance	farad	F	C/V
electric resistance	ohm	Ω	V/A
conductance	siemens	S	A/V
magnetic flux	weber	Wb	V·s
magnetic-flux density	tesla	T	Wb/m^2
inductance	henry	H	Wb/A
luminous flux	lumen	lm	cd·sr
illuminance	lux	lx	lm/m^2
activity (of radionuclides)	becquerel	Bq	l/s
absorbed dose	gray	Gy	J/kg

TABLE 1-2*b* Additional Common Derived Units of SI

Quantity	Unit	Symbol
acceleration	meter per second squared	m/s^2
angular acceleration	radian per second squared	rad/s^2
angular velocity	radian per second	rad/s
area	square meter	m^2
concentration (of amount of substance)	mole per cubic meter	mol/m^3
current density	ampere per square meter	A/m^2
density, mass	kilogram per cubic meter	kg/m^3
electric-charge density	coulomb per cubic meter	C/m^3
electric-field strength	volt per meter	V/m
electric-flux density	coulomb per square meter	C/m^2
energy density	joule per cubic meter	J/m^3
entropy	joule per kelvin	J/K
heat capacity	joule per kelvin	J/K
heat-flux density, irradiance	watt per square meter	W/m^2
luminance	candela per square meter	cd/m^2
magnetic-field strength	ampere per meter	A/m
molar energy	joule per mole	J/mol
molar entropy	joule per mole-kelvin	J/(mol·K)
molar-heat capacity	joule per mole-kelvin	J/(mol·K)
moment of force	newton-meter	N·m
permeability	henry per meter	H/m
permittivity	farad per meter	F/m
radiance	watt per square-meter-steradian	W/(m^2·sr)
radiant intensity	watt per steradian	W/sr
specific-heat capacity	joule per kilogram-kelvin	J/(kg·K)
specific energy	joule per kilogram	J/kg
specific entropy	joule per kilogram-kelvin	J/(kg·K)
specific volume	cubic meter per kilogram	m^3/kg
surface tension	newton per meter	N/m
thermal conductivity	watt per meter-kelvin	W/(m·K)
velocity	meter per second	m/s
viscosity, dynamic	pascal-second	Pa·s
viscosity, kinematic	square meter per second	m^2/s
volume	cubic meter	m^3
wave number	1 per meter	1/m

TABLE 1-3 SI Prefixes

Multiplication factor	Prefix	Symbol
1 000 000 000 000 000 000 = 10^{18}	exa	E
1 000 000 000 000 000 = 10^{15}	peta	P
1 000 000 000 000 = 10^{12}	tera	T
1 000 000 000 = 10^{9}	giga	G
1 000 000 = 10^{6}	mega	M
1 000 = 10^{3}	kilo	k
100 = 10^{2}	hecto*	h
10 = 10^{1}	deka*	da
0.1 = 10^{-1}	deci*	d
0.01 = 10^{-2}	centi	c
0.001 = 10^{-3}	milli	m
0.000 001 = 10^{-6}	micro	μ
0.000 000 001 = 10^{-9}	nano	n
0.000 000 000 001 = 10^{-12}	pico	p
0.000 000 000 000 001 = 10^{-15}	femto	f
0.000 000 000 000 000 001 = 10^{-18}	atto	a

*Generally to be avoided.

TABLE 1-4 Conversion Factors: U.S. Customary and Commonly Used Units to SI Units

Quantity	Customary or commonly used unit	SI unit	Alternate SI unit	Conversion factor; multiply customary unit by factor to obtain SI unit
	Space,† time			
Length	naut mi	km		1.852* E + 00
	mi	km		1.609 344* E + 00
	chain	m		2.011 68* E + 01
	link	m		2.011 68* E − 01
	fathom	m		1.828 8* E + 00
	yd	m		9.144* E − 01
	ft	m		3.048* E − 01
		cm		3.048* E + 01
	in	mm		2.54* E + 01
	in	cm		2.54 E + 00
	mil	μm		2.54* E + 01
Length/length	ft/mi	m/km		1.893 939 E − 01
Length/volume	ft/U.S. gal	m/m^3		8.051 964 E + 01
	ft/ft^3	m/m^3		1.076 391 E + 01
	ft/bbl	m/m^3		1.917 134 E + 00
Area	mi^2	km^2		2.589 988 E + 00
	section	ha		2.589 988 E + 02
	acre	ha		4.046 856 E − 01
	ha	m^2		1.000 000* E + 04
	yd^2	m^2		8.361 274 E − 01
	ft^2	m^2		9.290 304* E − 02
	in^2	mm^2		6.451 6* E + 02
		cm^2		6.451 6* E + 00
Area/volume	ft^2/in^3	m^2/cm^3		5.699 291 E − 03
	ft^2/ft^3	m^2/m^3		3.280 840 E + 00
Volume	cubem	km^3		4.168 182 E + 00
	acre·ft	m^3		1.233 482 E + 03
		ha·m		1.233 482 E − 01
	yd^3	m^3		7.645 549 E − 01
	bbl (42 U.S. gal)	m^3		1.589 873 E − 01
	ft^3	m^3		2.831 685 E − 02
		dm^3	L	2.831 685 E + 01
	U.K. gal	m^3		4.546 092 E − 03
		dm^3	L	4.546 092 E + 00
	U.S. gal	m^3		3.785 412 E − 03
		dm^3	L	3.785 412 E + 00
	U.K. qt	dm^3	L	1.136 523 E + 00
	U.S. qt	dm^3	L	9.463 529 E − 01
	U.S. pt	dm^3	L	4.731 765 E − 01
	U.K. fl oz	cm^3		2.841 307 E + 01
	U.S. fl oz	cm^3		2.957 353 E + 01
	in^3	cm^3		1.638 706 E + 01
Volume/length (linear displacement)	bbl/in	m^{3}m		6.259 342 E + 00
	bbl/ft	m^3/m		5.216 119 E − 01
	ft^3/ft	m^3/m		9.290 304* E − 02
	U.S. gal/ft	m^3/m		1.241 933 E − 02
		L/m		1.241 933 E + 01
Plane angle	rad	rad		1
	deg (°)	rad		1.745 329 E − 02
	min (′)	rad		2.908 882 E − 04
	sec (″)	rad		4.848 137 E − 06
Solid angle	sr	sr		1
Time	year	a		1
	week	d		7.0* E + 00
	h	s		3.6* E + 03
		min		6.0* E + 01
	min	s		6.0* E + 01
		h		1.666 667 E − 02
	mμs	ns		1
	Mass, amount of substance			
Mass	U.K. ton	Mg	t	1.016 047 E + 00
	U.S. ton	Mg	t	9.071 847 E − 01
	U.K. cwt	kg		5.080 234 E + 01
	U.S. cwt	kg		4.535 924 E + 01
	lbm	kg		4.535 924 E − 01
	oz (troy)	g		3.110 348 E + 01
	oz (av)	g		2.834 952 E + 01
	gr	mg		6.479 891 E + 01

TABLE 1-4 Conversion Factors: U.S. Customary and Commonly Used Units to SI Units (*Continued*)

Quantity	Customary or commonly used unit	SI unit	Alternate SI unit	Conversion factor; multiply customary unit by factor to obtain SI unit
Amount of substance	lbm·mol	kmol		4.535 924 E − 01
	std m^3(0°C, 1 atm)	kmol		4.461 58 E − 02
	std ft^3 (60°F, 1 atm)	kmol		1.195 30 E − 03
Enthalpy, calorific value, heat, entropy, heat capacity				
Calorific value, enthalpy (mass basis)	Btu/lbm	MJ/kg		2.326 000 E − 03
		kJ/kg	J/g	2.326 000 E + 00
		kWh/kg		6.461 112 E − 04
	cal/g	kJ/kg	J/g	4.184* E + 00
	cal/lbm	J/kg		9.224 141 E + 00
Caloric value, enthalpy (mole basis)	kcal/(g·mol)	kJ/kmol		4.184* E + 03
	Btu/(lb·mol)	kJ/kmol		2.326 000 E + 00
Calorific value (volume basis—solids and liquids)	Btu/U.S. gal	MJ/m^3	kJ/dm^3	2.787 163 E − 01
		kJ/m^3		2.787 163 E + 02
		kWh/m^3		7.742 119 E − 02
	Btu/U.K. gal	MJ/m^3	kJ/dm^3	2.320 800 E − 01
		kJ/m^3		2.320 800 E + 02
	Btu/ft^3	kWh/m^3		6.446 667 E − 02
		MJ/m^3	kJ/dm^3	3.725 895 E − 02
		kJ/m^3		3.725 895 E + 01
		kWh/m^3		1.034 971 E − 02
	cal/mL	MJ/m^3		4.184* E + 00
	(ft·lbf)/U.S. gal	kJ/m^3		3.581 692 E − 01
Calorific value (volume basis—gases)	cal/mL	kJ/m^3	J/dm^3	4.184* E + 03
	kcal/m^3	kJ/m^3	J/dm^3	4.184* E + 00
	Btu/ft^3	kJ/m^3	J/dm^3	3.725 895 E + 01
		kWh/m^3		1.034 971 E − 02
Specific entropy	Btu/(lbm·°R)	kJ/(kg·K)	J/(g·K)	4.186 8* E + 00
	cal/(g·K)	kJ/(kg·K)	J/(g·K)	4.184* E + 00
	kcal/(kg·°C)	kJ/(kg·K)	J/(g·K)	4.184* E + 00
Specific-heat capacity (mass basis)	kWh/(kg·°C)	kJ/(kg·K)	J/(g·K)	3.6* E + 03
	Btu/(lbm·°F)	kJ/(kg·K)	J/(g·K)	4.186 8* E + 00
	kcal/(kg·°C)	kJ/(kg·K)	J/(g·K)	4.184* E + 00
Specific-heat capacity (mole basis)	Btu/(lb·mol·°F)	kJ/(kmol·K)		4.186 8* E + 00
	cal/(g·mol·°C)	kJ/(kmol·K)		4.184* E + 00
Temperature, pressure, vacuum				
Temperature (absolute)	°R	K		5/9
	K	K		1
Temperature (traditional)	°F	°C		5/9(°F − 32)
Temperature (difference)	°F	K, °C		5/9
Pressure	atm (760 mmHg at 0°C or 14,696 psi)	MPa		1.013 250* E − 01
		kPa		1.013 250* E + 02
		bar		1.013 250* E + 00
	bar	MPa		1.0* E − 01
		kPa		1.0* E + 02
	mmHg (0°C) = torr	MPa		6.894 757 E − 03
		kPa		6.894 757 E + 00
		bar		6.894 757 E − 02
	μmHg (0°C)	kPa		3.376 85 E + 00
	μ bar	kPa		2.488 4 E − 01
	mmHg = torr (0°C)	kPa		1.333 224 E − 01
	cmH_2O (4°C)	kPa		9.806 38 E − 02
	lbf/ft^2 (psf)	kPa		4.788 026 E − 02
	mHg (0°C)	Pa		1.333 224 E − 01
	bar	Pa		1.0* E + 05
	dyn/cm^2	Pa		1.0* E − 01
Vacuum, draft	inHg (60°F)	kPa		3.376 85 E + 00
	inH_2O (39.2°F)	kPa		2.490 82 E − 01
	inH_2O (60°F)	kPa		2.488 4 E − 01
	mmHg (0°C) = torr	kPa		1.333 224 E − 01
	cmH_2O (4°C)	kPa		9.806 38 E − 02
Liquid head	ft	m		3.048* E − 01
	in	mm		2.54* E + 01
		cm		2.54* E + 00
Pressure drop/length	psi/ft	kPa/m		2.262 059 E + 01

TABLE 1-4 Conversion Factors: U.S. Customary and Commonly Used Units to SI Units *(Continued)*

Quantity	Customary or commonly used unit	SI unit	Alternate SI unit	Conversion factor; multiply customary unit by factor to obtain SI unit
	Density, specific volume, concentration, dosage			
Density	lbm/ft^3	kg/m^3		1.601 846 E + 01
		g/m^3		1.601 846 E + 04
	lbm/U.S. gal	kg/m^3		1.198 264 E + 02
		g/cm^3		1.198 264 E − 01
	lbm/U.K. gal	kg/m^3		9.977 633 E + 01
	lbm/ft^3	kg/m^3		1.601 846 E + 01
		g/cm^3		1.601 846 E − 02
	g/cm^3	kg/m^3		1.0* E + 03
	lbm/ft^3	kg/m^3		1.601 846 E + 01
Specific volume	ft^3/lbm	m^3/kg		6.242 796 E − 02
		m^3/g		6.242 796 E − 05
	ft^3/lbm	dm^3/kg		6.242 796 E + 01
	U.K. gal/lbm	dm^3/kg	cm^3/g	1.002 242 E + 01
	U.S. gal/lbm	dm^3/kg	cm^3/g	8.345 404 E + 00
Specific volume (mole basis)	L/(g·mol)	m^3/kmol		1
	ft^3/(lb·mol)	m^3/kmol		6.242 796 E − 02
Specific volume	bbl/U.S. ton	m^3/t		1.752 535 E − 01
	bbl/U.K. ton	m^3/t		1.564 763 E − 01
Yield	bbl/U.S. ton	dm^3/t	L/t	1.752 535 E + 02
	bbl/U.K. ton	dm^3/t	L/t	1.564 763 E + 02
	U.S. gal/U.S. ton	dm^3/t	L/t	4.172 702 E + 00
	U.S. gal/U.K. ton	dm^3/t	L/t	3.725 627 E + 00
Concentration (mass/mass)	wt %	kg/kg		1.0* E − 02
		g/kg		1.0* E + 01
	wt ppm	mg/kg		1
Concentration (mass/volume)	lbm/bbl	kg/m^3	g/dm^3	2.853 010 E + 00
	g/U.S. gal	kg/m^3		2.641 720 E − 01
	g/U.K. gal	kg/m^3	g/L	2.199 692 E − 01
	lbm/1000 U.S. gal	g/m^3	mg/dm^3	1.198 264 E + 02
	lbm/1000 U.K. gal	g/m^3	mg/dm^3	9.977 633 E + 01
	gr/U.S. gal	g/m^3	mg/dm^3	1.711 806 E + 01
	gr/ft^3	mg/m^3		2.288 351 E + 03
	lbm/1000 bbl	g/m^3	mg/dm^3	2.853 010 E + 00
	mg/U.S. gal	g/m^3	mg/dm^3	2.641 720 E − 01
	gr/100 ft^3	mg/m^3		2.288 351 E + 01
Concentration (volume/volume)	ft^3/ft^3	m^3/m^3		1
	bbl/(acre·ft)	m^3/m^3		1.288 931 E − 04
	vol%	m^3/m^3		1.0* E − 02
	U.K. gal/ft^3	dm^3/m^3	L/m^3	1.605 437 E + 02
	U.S. gal/ft^3	dm^3/m^3	L/m^3	1.336 806 E + 02
	mL/U.S. gal	dm^3/m^3	L/m^3	2.641 720 E − 01
	mL/U.K. gal	dm^3/m^3	L/m^3	2.199 692 E − 01
	vol ppm	cm^3/m^3		1
		dm^3/m^3	L/m^3	1.0* E − 03
	U.K. gal/1000 bbl	cm^3/m^3		2.859 403 E + 01
	U.S. gal/1000 bbl	cm^3/m^3		2.380 952 E + 01
	U.K. pt/1000 bbl	cm^3/m^3		3.574 253 E + 00
Concentration (mole/volume)	(lb·mol)/U.S. gal	kmol/m^3		1.198 264 E + 02
	(lb·mol)/U.K. gal	kmol/m^3		9.977 644 E + 01
	(lb·mol)/ft^3	kmol/m^3		1.601 846 E + 01
	std ft^3 (60°F, 1 atm)/bbl	kmol/m^3		7.518 21 E − 03
Concentration (volume/mole)	U.S. gal/1000 std ft^3 (60°F/60°F)	dm^3/kmol	L/kmol	3.166 91 E + 00
	bbl/million std ft^3 (60°F/60°F)	dm^3/kmol	L/kmol	1.330 10 E − 01
	Facility throughput, capacity			
Throughput (mass basis)	U.K. ton/year	t/a		1.016 047 E + 00
	U.S. ton/year	t/a		9.071 847 E − 01
	U.K. ton/day	t/d		1.016 047 E + 00
		t/h		4.233 529 E − 02
	U.S. ton/day	t/d		9.071 847 E − 01
		t/h		3.779 936 E − 02
	U.K. ton/h	t/h		1.016 047 E + 00
	U.S. ton/h	t/h		9.071 847 E − 01
	lbm/h	kg/h		4.535 924 E − 01

TABLE 1-4 Conversion Factors: U.S. Customary and Commonly Used Units to SI Units *(Continued)*

Quantity	Customary or commonly used unit	SI unit	Alternate SI unit	Conversion factor; multiply customary unit by factor to obtain SI unit
Throughput (volume basis)	bbl/day	t/a		5.803 036 E + 01
		m^3/d		1.589 873 E − 01
	ft^3/day	m^3/h		1.179 869 E − 03
	bbl/h	m^3/h		1.589 873 E − 01
	ft^3/h	m^3/h		2.831 685 E − 02
	U.K. gal/h	m^3/h		4.546 092 E − 03
		L/s		1.262 803 E − 03
	U.S. gal/h	m^3/h		3.785 412 E − 03
		L/s		1.051 503 E − 03
	U.K. gal/min	m^3/h		2.727 655 E − 01
		L/s		7.576 819 E − 02
	U.S. gal/min	m^3/h		2.271 247 E − 01
		L/s		6.309 020 E − 02
Throughput (mole basis)	(lbm·mol)/h	kmol/h		4.535 924 E − 01
		kmol/s		1.259 979 E − 04
	Flow rate			
Flow rate (mass basis)	U.K. ton/min	kg/s		1.693 412 E + 01
	U.S. ton/min	kg/s		1.511 974 E + 01
	U.K. ton/h	kg/s		2.822 353 E − 01
	U.S. ton/h	kg/s		2.519 958 E − 01
	U.K. ton/day	kg/s		1.175 980 E − 02
	U.S. ton/day	kg/s		1.049 982 E − 02
	million lbm/year	kg/s		5.249 912 E + 00
	U.K. ton/year	kg/s		3.221 864 E − 05
	U.S. ton/year	kg/s		2.876 664 E − 05
	lbm/s	kg/s		4.535 924 E − 01
	lbm/min	kg/s		7.559 873 E − 03
	lbm/h	kg/s		1.259 979 E − 04
Flow rate (volume basis)	bbl/day	m^3/d		1.589 873 E − 01
		L/s		1.840 131 E − 03
	ft^3/day	m^3/d		2.831 685 E − 02
		L/s		3.277 413 E − 04
	bbl/h	m^3/s		4.416 314 E − 05
		L/s		4.416 314 E − 02
	ft^3/h	m^3/s		7.865 791 E − 06
		L/s		7.865 791 E − 03
	U.K. gal/h	dm^3/s	L/s	1.262 803 E − 03
	U.S. gal/h	dm^3/s	L/s	1.051 503 E − 03
	U.K. gal/min	dm^3/s	L/s	7.576 820 E − 02
	U.S. gal/min	dm^3/s	L/s	6.309 020 E − 02
	ft^3/min	dm^3/s	L/s	4.719 474 E − 01
	ft^3/s	dm^3/s	L/s	2.831 685 E + 01
Flow rate (mole basis)	(lb·mol)/s	kmol/s		4.535 924 E − 01
	(lb·mol)/h	kmol/s		1.259 979 E − 04
	million scf/D	kmol/s		1.383 45 E − 02
Flow rate/length (mass basis)	lbm/(s·ft)	kg/(s·m)		1.488 164 E + 00
	lbm/(h·ft)	kg/(s·m)		4.133 789 E − 04
Flow rate/length (volume basis)	U.K. gal/(min·ft)	m^2/s	m^3/(s·m)	2.485 833 E − 04
	U.S. gal/(min·ft)	m^2/s	m^3/(s·m)	2.069 888 E − 04
	U.K. gal/(h·in)	m^2/s	m^3/(s·m)	4.971 667 E − 05
	U.S. gal/(h·in)	m^2/s	m^3/(s·m)	4.139 776 E − 05
	U.K. gal/(h·ft)	m^2/s	m^3/(s·m)	4.143 055 E − 06
	U.S. gal/(h·ft)	m^2/s	m^3/(s·m)	3.449 814 E − 06
Flow rate/area (mass basis)	lbm/(s·ft^2)	kg/(s·m^2)		4.882 428 E + 00
	lbm/(h·ft^2)	kg/(s·m^2)		1.356 230 E − 03
Flow rate/area (volume basis)	ft^3/(s·ft^2)	m/s	m^3/(s·m^2)	3.048* E − 01
	ft^3/(min·ft^2)	m/s	m^3/(s·m^2)	5.08* E − 03
	U.K. gal/(h·in^2)	m/s	m^3/(s·m^2)	1.957 349 E − 03
	U.S. gal/(h·in^2)	m/s	m^3/(s·m^2)	1.629 833 E − 03
	U.K. gal/(min·ft^2)	m/s	m^3/(s·m^2)	8.155 621 E − 04
	U.S. gal/(min·ft^2)	m/s	m^3/(s·m^2)	6.790 972 E − 04
	U.K. gal/(h·ft^2)	m/s	m^3/(s·m^2)	1.359 270 E − 05
	U.S. gal/(h·ft^2)	m/s	m^3/(s·m^2)	1.131 829 E − 05

TABLE 1-4 Conversion Factors: U.S. Customary and Commonly Used Units to SI Units *(Continued)*

Quantity	Customary or commonly used unit	SI unit	Alternate SI unit	Conversion factor; multiply customary unit by factor to obtain SI unit
	Energy, work, power			
Energy, work	therm	MJ		1.055 056 E + 02
		kJ		1.055 056 E + 05
		kWh		2.930 711 E + 01
	U.S. tonf·mi	MJ		1.431 744 E + 01
	hp·h	MJ		2.684 520 E + 00
		kJ		2.684 520 E + 03
		kWh		7.456 999 E − 01
	ch·h or CV·h	MJ		2.647 780 E + 00
		kJ		2.647 780 E + 03
		kWh		7.354 999 E − 01
	kWh	MJ		3.6° E + 00
		kJ		3.6° E + 03
	Chu	kJ		1.899 101 E + 00
		kWh		5.275 280 E − 04
	Btu	kJ		1.055 056 E + 00
		kWh		2.930 711 E − 04
	kcal	kJ		4.184° E + 00
	cal	kJ		4.184° E − 03
	ft·lbf	kJ		1.355 818 E − 03
	lbf·ft	kJ		1.355 818 E − 03
	J	kJ		1.0° E − 03
	$(lbf{\cdot}ft^2)/s^2$	kJ		4.214 011 E − 05
	erg	J		1.0° E − 07
Impact energy	kgf·m	J		9.806 650° E + 00
	lbf·ft	J		1.355 818 E + 00
Surface energy	erg/cm^2	mJ/m^2		1.0° E + 00
Specific-impact energy	$(kgf{\cdot}m)/cm^2$	J/cm^2		9.806 650° E − 02
	$(lbf{\cdot}ft)/in^2$	J/cm^2		2.101 522 E − 03
Power	million Btu/h	MW		2.930 711 E − 01
	ton of refrigeration	kW		3.516 853 E + 00
	Btu/s	kW		1.055 056 E + 00
	kW	kW		1
	hydraulic horsepower—hhp	kW		7.460 43 E − 01
	hp (electric)	kW		7.46° E − 01
	hp [(550 ft·lbf)/s]	kW		7.456 999 E − 01
	ch or CV	kW		7.354 999 E − 01
	Btu/min	kW		1.758 427 E − 02
	(ft·lbf)/s	kW		1.355 818 E − 03
	kcal/h	W		1.162 222 E + 00
	Btu/h	W		2.930 711 E − 01
	(ft·lbf)/min	W		2.259 697 E − 02
Power/area	$Btu/(s{\cdot}ft^2)$	kW/m^2		1.135 653 E + 01
	$cal/(h{\cdot}cm^2)$	kW/m^2		1.162 222 E − 02
	$Btu/(h{\cdot}ft^2)$	kW/m^2		3.154 591 E − 03
Heat-release rate, mixing power	hp/ft^3	kW/m^3		2.633 414 E + 01
	$cal/(h{\cdot}cm^3)$	kW/m^3		1.162 222 E + 00
	$Btu/(s{\cdot}ft^3)$	kW/m^3		3.725 895 E + 01
	$Btu/(h{\cdot}ft^3)$	kW/m^3		1.034 971 E − 02
Cooling duty (machinery)	Btu/(bhp·h)	W/kW		3.930 148 E − 01
Specific fuel consumption (mass basis)	lbm/(hp·h)	mg/J	kg/MJ	1.689 659 E − 01
		kg/kWh		6.082 774 E − 01
Specific fuel consumption (volume basis)	m^3/kWh	dm^3/MJ	mm^3/J	2.777 778 E + 02
	U.S. gal/(hp·h)	dm^3/MJ	mm^3/J	1.410 089 E + 00
	U.K. pt/(hp·h)	dm^3/MJ	mm^3/J	2.116 806 E − 01
Fuel consumption	U.K. gal/mi	$dm^3/100$ km	L/100 km	2.824 807 E + 02
	U.S. gal/mi	$dm^3/100$ km	L/100 km	2.352 146 E + 02
	mi/U.S. gal	km/dm^3	km/L	4.251 437 E − 01
	mi/U.K. gal	km/dm^3	km/L	3.540 064 E − 01

TABLE 1-4 Conversion Factors: U.S. Customary and Commonly Used Units to SI Units *(Continued)*

Quantity	Customary or commonly used unit	SI unit	Alternate SI unit	Conversion factor; multiply customary unit by factor to obtain SI unit
Velocity (linear), speed	knot	km/h		1.852* E + 00
	mi/h	km/h		1.609 344* E + 00
	ft/s	m/s		3.048* E − 01
		cm/s		3.048* E + 01
	ft/min	m/s		5.08* E − 03
	ft/h	mm/s		8.466 667 E − 02
	ft/day	mm/s		3.527 778 E − 03
		m/d		3.048* E − 01
	in/s	mm/s		2.54* E + 01
	in/min	mm/s		4.233 333 E − 01
Corrosion rate	in/year (ipy)	mm/a		2.54* E + 01
	mil/year	mm/a		2.54* E − 02
Rotational frequency	r/min	r/s		1.666 667 E − 02
		rad/s		1.047 198 E − 01
Acceleration (linear)	ft/s^2	m/s^2		3.048* E − 01
		cm/s^2		3.048* E + 01
Acceleration (rotational)	rpm/s	rad/s^2		1.047 198 E − 01
Momentum	(lbm·ft)/s	(kg·m)/s		1.382 550 E − 01
Force	U.K. tonf	kN		9.964 016 E + 00
	U.S. tonf	kN		8.896 443 E + 00
	kgf (kp)	N		9.806 650* E + 00
	lbf	N		4.448 222 E + 00
	dyn	mN		1.0 E − 02
Bending moment, torque	U.S. tonf·ft	kN·m		2.711 636 E + 00
	kgf·m	N·m		9.806 650* E + 00
	lbf·ft	N·m		1.355 818 E + 00
	lbf·in	N·m		1.129 848 E − 01
Bending moment/length	(lbf·ft)/in	(N·m)/m		5.337 866 E + 01
	(lbf·in)/in	(N·m)/m		4.448 222 E + 00
Moment of inertia	lbm·ft^2	kg·m^2		4.214 011 E − 02
Stress	U.S. tonf/in^2	MPa	N/mm^2	1.378 951 E + 01
	kgf/mm^2	MPa	N/mm^2	9.806 650* E + 00
	U.S. tonf/ft^2	MPa	N/mm^2	9.576 052 E − 02
	lbf/in^2 (psi)	MPa	N/mm^2	6.894 757 E − 03
	lbf/ft^2 (psf)	kPa		4.788 026 E − 02
	dyn/cm^2	Pa		1.0* E − 01
Mass/length	lbm/ft	kg/m		1.488 164 E + 00
Mass/area structural loading, bearing capacity (mass basis)	U.S. ton/ft^2	Mg/m^2		9.764 855 E + 00
	lbm/ft^2	kg/m^2		4.882 428 E + 00
	Miscellaneous transport properties			
Diffusivity	ft^2/s	m^2/s		9.290 304* E − 02
	m^2/s	mm^2/s		1.0* E + 06
	ft^2/h	m^2/s		2.580 64* E − 05
Thermal resistance	(°C·m^2·h)/kcal	(K·m^2)/kW		8.604 208 E + 02
	(°F·ft^2·h)/Btu	(K·m^2)/kW		1.761 102 E + 02
Heat flux	Btu/(h·ft^2)	kW/m^2		3.154 591 E − 03
Thermal conductivity	(cal·cm)/(s·cm^2·°C)	W/(m·K)		4.184* E + 02
	(Btu·ft)/(h·ft^2·°F)	W/(m·K)		1.730 735 E + 00
		(kJ·m)/(h·m^2·K)		6.230 646 E + 00
	(kcal·m)/(h·m^2·°C)	W/(m·K)		1.162 222 E + 00
	(Btu·in)/(h·ft^2·°F)	W/(m·K)		1.442 279 E − 01
	(cal·cm)/(h·cm^2·°C)	W/(m·K)		1.162 222 E − 01
Heat-transfer coefficient	cal/(s·cm^2·°C)	kW/(m^2·K)		4.184* E + 01
	Btu/(s·ft^2·°F)	kW/(m^2·K)		2.044 175 E + 01
	cal/(h·cm^2·°C)	kW/(m^2·K)		1.162 222 E − 02
	Btu/(h·ft^2·°F)	kW/(m^2·K)		5.678 263 E − 03
		kJ/(h·m^2·K)		2.044 175 E + 01
	Btu/(h·ft^2·°R)	kW/(m^2·K)		5.678 263 E − 03
	kcal/(h·m^2·°C)	kW/(m^2·K)		1.162 222 E − 03

TABLE 1-4 Conversion Factors: U.S. Customary and Commonly Used Units to SI Units *(Continued)*

Quantity	Customary or commonly used unit	SI unit	Alternate SI unit	Conversion factor; multiply customary unit by factor to obtain SI unit
Volumetric heat-transfer coefficient	Btu/(s·ft³·°F)	kW/(m³·K)		6.706 611 E + 01
	Btu/(h·ft³·°F)	kW/(m³·K)		1.862 947 E − 02
Surface tension	dyn/cm	mN/m		1
Viscosity (dynamic)	(lbf·s)/in²	Pa·s	(N·s)/m²	6.894 757 E + 03
	(lbf·s)/ft²	Pa·s	(N·s)/m²	4.788 026 E + 01
	(kgf·s)/m²	Pa·s	(N·s)/m²	9.806 650* E + 00
	lbm/(ft·s)	Pa·s	(N·s)/m²	1.488 164 E + 00
	(dyn·s)/cm²	Pa·s	(N·s)/m²	1.0* E − 01
	cP	Pa·s	(N·s)/m²	1.0* E − 03
	lbm/(ft·h)	Pa·s	(N·s)/m²	4.133 789 E − 04
Viscosity (kinematic)	ft²/s	m²/s		9.290 304* E − 02
	in²/s	mm²/s		6.451 6* E + 02
	m²/h	mm²/s		2.777 778 E + 02
	ft²/h	m²/s		2.580 64* E − 05
	cSt	mm²/s		1
Permeability	darcy	μm²		9.869 233 E − 01
	millidarcy	μm²		9.869 233 E − 04
Thermal flux	Btu/(h·ft²)	W/m²		3.152 E + 00
	Btu/(s·ft²)	W/m²		1.135 E + 04
	cal/(s·cm²)	W/m²		4.184 E + 04
Mass-transfer coefficient	(lb·mol)/[h·ft²(lb·mol/ft³)]	m/s		8.467 E − 05
	(g·mol)/[s·m²(g·mol/L)]	m/s		1.0 E + 01
	Electricity, magnetism			
Admittance	S	S		1
Capacitance	μF	μF		1
Charge density	C/mm³	C/mm³		1
Conductance	S	S		1
	℧ (mho)	S		1
Conductivity	S/m	S/m		1
	℧/m	S/m		1
	m℧/m	mS/m		1
Current density	A/mm²	A/mm²		1
Displacement	C/cm²	C/cm²		1
Electric charge	C	C		1
Electric current	A	A		1
Electric-dipole moment	C·m	C·m		1
Electric-field strength	V/m	V/m		1
Electric flux	C	C		1
Electric polarization	C/cm²	C/cm²		1
Electric potential	V	V		1
	mV	mV		1
Electromagnetic moment	A·m²	A·m²		1
Electromotive force	V	V		1
Flux of displacement	C	C		1
Frequency	cycles/s	Hz		1
Impedance	Ω	Ω		1
Linear-current density	A/mm	A/mm		1
Magnetic-dipole moment	Wb·m	Wb·m		1
Magnetic-field strength	A/mm	A/mm		1
	Oe	A/m		7.957 747 E + 01
	gamma	A/m		7.957 747 E − 04
Magnetic flux	mWb	mWb		1

TABLE 1-4 Conversion Factors: U.S. Customary and Commonly Used Units to SI Units *(Continued)*

Quantity	Customary or commonly used unit	SI unit	Alternate SI unit	Conversion factor; multiply customary unit by factor to obtain SI unit	
Magnetic-flux density	mT G gamma	mT T nT		1 1.0* 1	 E − 04
Magnetic induction	mT	mT		1	
Magnetic moment	A·m²	A·m²		1	
Magnetic polarization	mT	mT		1	
Magnetic potential difference	A	A		1	
Magnetic-vector potential	Wb/mm	Wb/mm		1	
Magnetization	A/mm	A/mm		1	
Modulus of admittance	S	S		1	
Modulus of impedance	Ω	Ω		1	
Mutual inductance	H	H		1	
Permeability	μH/m	μH/m		1	
Permeance	H	H		1	
Permittivity	μF/m	μF/m		1	
Potential difference	V	V		1	
Quantity of electricity	C	C		1	
Reactance	Ω	Ω		1	
Reluctance	H^{-1}	H^{-1}		1	
Resistance	Ω	Ω		1	
Resistivity	Ω·cm Ω·m	Ω·cm Ω·m		1 1	
Self-inductance	mH	mH		1	
Surface density of change	mC/m²	mC/m²		1	
Susceptance	S	S		1	
Volume density of charge	C/mm³	C/mm³		1	
	Acoustics, light, radiation				
Absorbed dose	rad	Gy		1.0*	E − 02
Acoustical energy	J	J		1	
Acoustical intensity	W/cm²	W/m²		1.0*	E + 04
Acoustical power	W	W		1	
Sound pressure	N/m²	N/m²		1.0*	
Illuminance	fc	lx		1.076 391	E + 01
Illumination	fc	lx		1.076 391	E + 01
Irradiance	W/m²	W/m²		1	
Light exposure	fc·s	lx·s		1.076 391	E + 01
Luminance	cd/m²	cd/m²		1	
Luminous efficacy	lm/W	lm/W		1	
Luminous exitance	lm/m²	lm/m²		1	
Luminous flux	lm	lm		1	
Luminous intensity	cd	cd		1	
Radiance	W/m²·sr	W/m²·sr		1	
Radiant energy	J	J		1	
Radiant flux	W	W		1	
Radiant intensity	W/sr	W/sr		1	
Radiant power	W	W		1	

TABLE 1-4 Conversion Factors: U.S. Customary and Commonly Used Units to SI Units *(Concluded)*

Quantity	Customary or commonly used unit	SI unit	Alternate SI unit	Conversion factor; multiply customary unit by factor to obtain SI unit	
Wavelength	Å	nm		1.0*	E − 01
Capture unit	10^{-3} cm^{-1}	m^{-1}		1.0*	E + 01
			10^{-3} cm^{-1}	1	
	m^{-1}	m^{-1}		1	
Radioactivity	Ci	Bq		3.7*	E + 10

*An asterisk indicates that the conversion factor is exact.

†Conversion factors for length, area, and volume are based on the international foot. The international foot is longer by 2 parts in 1 million than the U.S. Survey foot (land-measurement use).

NOTE: The following unit symbols are used in the table:

Unit symbol	Name	Unit symbol	Name
A	ampere	lm	lumen
a	annum (year)	lx	lux
Bq	becquerel	m	meter
C	coulomb	min	minute
cd	candela	′	minute
Ci	curie	N	newton
d	day	naut mi	U.S. nautical mile
°C	degree Celsius	Oe	oersted
°	degree	Ω	ohm
dyn	dyne	Pa	pascal
F	farad	rad	radian
fc	footcandle	r	revolution
G	gauss	S	siemens
g	gram	s	second
gr	grain	″	second
Gy	gray	sr	steradian
H	henry	St	stokes
h	hour	T	tesla
ha	hectare	t	tonne
Hz	hertz	V	volt
J	joule	W	watt
K	kelvin	Wb	weber
L, ℓ, l	liter		

TABLE 1-5 Metric Conversion Factors as Exact Numerical Multiples of SI Units

The first two digits of each numerical entry represent a power of 10. For example, the entry "–02 2.54" expresses the fact that 1 in = 2.54×10^{-2} m.

To convert from	To	Multiply by
abampere	ampere	+01 1.00
abcoulomb	coulomb	+01 1.00
abfarad	farad	+09 1.00
abhenry	henry	–09 1.00
abmho	mho	+09 1.00
abohm	ohm	–09 1.00
abvolt	volt	–08 1.00
acre	meter^2	+03 4.046 856
ampere (international of 1948)	ampere	–01 9.998 35
angstrom	meter	–10 1.00
are	meter^2	+02 1.00
astronomical unit	meter	+11 1.495 978
atmosphere	newton/meter^2	+05 1.013 25
bar	newton/meter^2	+05 1.00
barn	meter^2	–28 1.00
barrel (petroleum 42 gal)	meter^3	–01 1.589 873
barye	newton/meter^2	–01 1.00
British thermal unit (ISO/TC 12)	joule	+03 1.055 06
British thermal unit (International Steam Table)	joule	+03 1.055 04
British thermal unit (mean)	joule	+03 1.055 87
British thermal unit (thermochemical)	joule	+03 1.054 350
British thermal unit (39°F)	joule	+03 1.059 67
British thermal unit (60°F)	joule	+03 1.054 68
bushel (U.S.)	meter^3	–02 3.523 907
cable	meter	+02 2.194 56
caliber	meter	–04 2.54
calorie (International Steam Table)	joule	+00 4.1868
calorie (mean)	joule	+00 4.190 02
calorie (thermochemical)	joule	+00 4.184
calorie (15°C)	joule	+00 4.185 80
calorie (20°C)	joule	+00 4.181 90
calorie (kilogram, International Steam Table)	joule	+03 4.186 8
calorie (kilogram, mean)	joule	+03 4.190 02
calorie (kilogram, thermochemical)	joule	+03 4.184
carat (metric)	kilogram	–04 2.00
Celsius (temperature)	kelvin	$t_K = t_c + 273.15$
centimeter of mercury (0°C)	newton/meter^2	+03 1.333 22
centimeter of water (4°C)	newton/meter^2	+01 9.806 38
chain (engineer's)	meter	+01 3.048
chain (surveyor's or Gunter's)	meter	+01 2.011 68
circular mil	meter^2	–10 5.067 074
cord	meter^3	+00 3.624 556
coulomb (international of 1948)	coulomb	–01 9.998 35
cubit	meter	–01 4.572
cup	meter^3	–04 2.365 882
curie	disintegration/second	+10 3.70
day (mean solar)	second (mean solar)	+04 8.64
day (sidereal)	second (mean solar)	+04 8.616 409
degree (angle)	radian	–02 1.745 329
denier (international)	kilogram/meter	–07 1.111 111
dram (avoirdupois)	kilogram	–03 1.771 845
dram (troy or apothecary)	kilogram	–03 3.887 934
dram (U.S. fluid)	meter^3	–06 3.696 691
dyne	newton	–05 1.00
electron volt	joule	–19 1.602 10
erg	joule	–07 1.00
Fahrenheit (temperature)	kelvin	$t_K = (5/9)(t_F + 459.67)$
Fahrenheit (temperature)	Celsius	$t_c = (5/9)(t_F - 32)$
farad (international of 1948)	farad	–01 9.995 05
faraday (based on carbon 12)	coulomb	+04 9.648 70
faraday (chemical)	coulomb	+04 9.649 57
faraday (physical)	coulomb	+04 9.652 19
fathom	meter	+00 1.828 8
fermi (femtometer)	meter	–15 1.00
fluid ounce (U.S.)	meter^3	–05 2.957 352
foot	meter	–01 3.048
foot (U.S. survey)	meter	–01 3.048 006
foot of water (39.2°F)	newton/meter^2	+03 2.988 98
footcandle	lumen/meter^2	+01 1.076 391
footlambert	candela/meter^2	+00 3.426 259
furlong	meter	+02 2.011 68
gal (galileo)	meter/second^2	–02 1.00
gallon (U.K. liquid)	meter^3	–03 4.546 087
gallon (U.S. dry)	meter^3	–03 4.404 883
gallon (U.S. liquid)	meter^3	–03 3.785 411
gamma	tesla	–09 1.00
gauss	tesla	–04 1.00
gilbert	ampere turn	–01 7.957 747
gill (U.K.)	meter^3	–04 1.420 652
gill (U.S.)	meter^3	–04 1.182 941
grad	degree (angular)	–01 9.00
grad	radian	–02 1.570 796
grain	kilogram	–05 6.479 891
gram	kilogram	–03 1.00
hand	meter	–01 1.016
hectare	meter^2	+04 1.00
henry (international of 1948)	henry	+00 1.000 495
hogshead (U.S.)	meter^3	–01 2.384 809
horsepower (550 ft lbf/s)	watt	+02 7.456 998
horsepower (boiler)	watt	+03 9.809 50
horsepower (electric)	watt	+02 7.46
horsepower (metric)	watt	+02 7.354 99
horsepower (U.K.)	watt	+02 7.457
horsepower (water)	watt	+02 7.460 43
hour (mean solar)	second (mean solar)	+03 3.60
hour (sidereal)	second (mean solar)	+03 3.590 170
hundredweight (long)	kilogram	+01 5.080 234
hundredweight (short)	kilogram	+01 4.535 923
inch	meter	–02 2.54
inch of mercury (32°F)	newton/meter^2	+03 3.386 389
inch of mercury (60°F)	newton/meter^2	+03 3.376 85
inch of water (39.2°F)	newton/meter^2	+02 2.490 82
inch of water (60°F)	newton/meter^2	+02 2.4884
joule (international of 1948)	joule	+00 1.000 165
kayser	1/meter	+02 1.00
kilocalorie (International Steam Table)	joule	+03 4.186 74
kilocalorie (mean)	joule	+03 4.190 02
kilocalorie (thermochemical)	joule	+03 4.184
kilogram mass	kilogram	+00 1.00
kilogram-force (kgf)	newton	+00 9.806 65
kilopond-force	newton	+00 9.806 65
kip	newton	+03 4.448 221
knot (international)	meter/second	–01 5.144 444
lambert	candela/meter^2	+04 $1/\pi$
lambert	candela/meter^2	+03 3.183 098
langley	joule/meter^2	+04 4.184
lbf (pound-force, avoirdupois)	newton	+00 4.448 221
lbm (pound-mass, avoirdupois)	kilogram	–01 4.535 923
league (British nautical)	meter	+03 5.559 552
league (international nautical)	meter	+03 5.556
league (statute)	meter	+03 4.828 032
light-year	meter	+15 9.460 55
link (engineer's)	meter	–01 3.048
link (surveyor's or Gunter's)	meter	–01 2.011 68
liter	meter^3	–03 1.00
lux	lumen/meter^2	+00 1.00
maxwell	weber	–08 1.00
meter	wavelengths Kr 86	+06 1.650 763
micrometer	meter	–06 1.00
mil	meter	–05 2.54
mile (U.S. statute)	meter	+03 1.609 344
mile (U.K. nautical)	meter	+03 1.853 184
mile (international nautical)	meter	+03 1.852
mile (U.S. nautical)	meter	+03 1.852
millibar	newton/meter^2	+02 1.00
millimeter of mercury (0°C)	newton/meter^2	+02 1.333 224

TABLE 1-5 Metric Conversion Factors as Exact Numerical Multiples of SI Units *(Concluded)*

The first two digits of each numerical entry represent a power of 10. For example, the entry "−02 2.54" expresses the fact that 1 in = 2.54×10^{-2}.

To convert from	To	Multiply by
minute (angle)	radian	−04 2.908 882
minute (mean solar)	second (mean solar)	+01 6.00
minute (sidereal)	second (mean solar)	+01 5.983 617
month (mean calendar)	second (mean solar)	+06 2.628
nautical mile (international)	meter	+03 1.852
nautical mile (U.S.)	meter	+03 1.852
nautical mile (U.K.)	meter	+03 1.853 184
oersted	ampere/meter	+01 7.957 747
ohm (international of 1948)	ohm	+00 1.000 495
ounce-force (avoirdupois)	newton	−01 2.780 138
ounce-mass (avoirdupois)	kilogram	−02 2.834 952
ounce-mass (troy or apothecary)	kilogram	−02 3.110 347
ounce (U.S. fluid)	$meter^3$	−05 2.957 352
pace	meter	−01 7.62
parsec	meter	+16 3.083 74
pascal	$newton/meter^2$	+00 1.00
peck (U.S.)	$meter^3$	−03 8.809 767
pennyweight	kilogram	−03 1.555 173
perch	meter	+00 5.0292
phot	$lumen/meter^2$	+04 1.00
pica (printer's)	meter	−03 4.217 517
pint (U.S. dry)	$meter^3$	−04 5.506 104
pint (U.S. liquid)	$meter^3$	−04 4.731 764
point (printer's)	meter	−04 3.514 598
poise	$(newton\text{-}second)/meter^2$	−01 1.00
pole	meter	+00 5.0292
pound-force (lbf avoirdupois)	newton	+00 4.448 221
pound-mass (lbm avoirdupois)	kilogram	−01 4.535 923
pound-mass (troy or apothecary)	kilogram	−01 3.732 417
poundal	newton	−01 1.382 549
quart (U.S. dry)	$meter^3$	−03 1.101 220
quart (U.S. liquid)	$meter^3$	−04 9.463 529
rad (radiation dose absorbed)	joule/kilogram	−02 1.00
Rankine (temperature)	kelvin	$t_K = (5/9)t_R$
rayleigh (rate of photon emission)	$1/second\text{-}meter^2$	+10 1.00
rhe	$meter^2/(newton\text{-}second)$	+01 1.00
rod	meter	+00 5.0292
roentgen	coulomb/kilogram	−04 2.579 76
rutherford	disintegration/second	+06 1.00
second (angle)	radian	−06 4.848 136

To convert from	To	Multiply by
second (ephemeris)	second	+00 1.000 000
second (mean solar)	second (ephemeris)	Consult American Ephemeris and Nautical Almanac
second (sidereal)	second (mean solar)	−01 9.972 695
section	$meter^2$	+06 2.589 988
scruple (apothecary)	kilogram	−03 1.295 978
shake	second	−08 1.00
skein	meter	+02 1.097 28
slug	kilogram	+01 1.459 390
span	meter	−01 2.286
statampere	ampere	−10 3.335 640
statcoulomb	coulomb	−10 3.335 640
statfarad	farad	−12 1.112 650
stathenry	henry	+11 8.987 554
statmho	mho	−12 1.112 650
statohm	ohm	+11 8.987 554
statute mile (U.S.)	meter	+03 1.609 344
statvolt	volt	+02 2.997 925
stere	$meter^3$	+00 1.00
stilb	$candela/meter^2$	+04 1.00
stoke	$meter^2/second$	−04 1.00
tablespoon	$meter^3$	−05 1.478 676
teaspoon	$meter^3$	−06 4.928 921
ton (assay)	kilogram	−02 2.916 666
ton (long)	kilogram	+03 1.016 046
ton (metric)	kilogram	+03 1.00
ton (nuclear equivalent of TNT)	joule	+09 4.20
ton (register)	$meter^3$	+00 2.831 684
ton (short, 2000 lb)	kilogram	+02 9.071 847
tonne	kilogram	+03 1.00
torr (0°C)	$newton/meter^2$	+02 1.333 22
township	$meter^2$	+07 9.323 957
unit pole	weber	−07 1.256 637
volt (international of 1948)	volt	+00 1.000 330
watt (international of 1948)	watt	+00 1.000 165
yard	meter	−01 9.144
year (calendar)	second (mean solar)	+07 3.1536
year (sidereal)	second (mean solar)	+07 3.155 815
year (tropical)	second (mean solar)	+07 3.155 692
year 1900, tropical, Jan., day 0, hour 12	second (ephemeris)	+07 3.155 692
year 1900, tropical, Jan., day 0, hour 12	second	+07 3.155 692

TABLE 1-6 Alphabetical Listing of Common Conversions

To convert from	To	Multiply by
Acres	Square feet	43,560
Acres	Square meters	4074
Acres	Square miles	0.001563
Acre-feet	Cubic meters	1233
Ampere-hours (absolute)	Coulombs (absolute)	3600
Angstrom units	Inches	3.937×10^{-9}
Angstrom units	Meters	1×10^{-10}
Angstrom units	Microns	1×10^{-4}
Atmospheres	Millimeters of mercury at 32°F	760
Atmospheres	Dynes per square centimeter	1.0133×10^{6}
Atmospheres	Newtons per square meter	101,325
Atmospheres	Feet of water at 39.1°F	33.90
Atmospheres	Grams per square centimeter	1033.3
Atmospheres	Inches of mercury at 32°F	29.921
Atmospheres	Pounds per square foot	2116.3
Atmospheres	Pounds per square inch	14.696
Bags (cement)	Pounds (cement)	94
Barrels (cement)	Pounds (cement)	376
Barrels (oil)	Cubic meters	0.15899
Barrels (oil)	Gallons	42
Barrels (U.S. liquid)	Cubic meters	0.11924
Barrels (U.S. liquid)	Gallons	31.5
Barrels per day	Gallons per minute	0.02917
Bars	Atmospheres	0.9869
Bars	Newtons per square meter	1×10^{5}
Bars	Pounds per square inch	14.504
Board feet	Cubic feet	1/12
Boiler horsepower	B.t.u. per hour	33,480
Boiler horsepower	Kilowatts	9.803
B.t.u.	Calories (gram)	252
B.t.u.	Centigrade heat units (c.h.u. or p.c.u.)	0.55556
B.t.u.	Foot-pounds	777.9
B.t.u.	Horsepower-hours	3.929×10^{-4}
B.t.u.	Joules	1055.1
B.t.u.	Liter-atmospheres	10.41
B.t.u.	Pounds carbon to CO_2	6.88×10^{-5}
B.t.u.	Pounds water evaporated from and at 212°F	0.001036
B.t.u.	Cubic foot-atmospheres	0.3676
B.t.u.	Kilowatt-hours	2.930×10^{-4}
B.t.u. per cubic foot	Joules per cubic meter	37,260
B.t.u. per hour	Watts	0.29307
B.t.u. per minute	Horsepower	0.02357
B.t.u. per pound	Joules per kilogram	2326
B.t.u. per pound per degree Fahrenheit	Calories per gram per degree centigrade	1
B.t.u. per pound per degree Fahrenheit	Joules per kilogram per degree Kelvin	4186.8
B.t.u. per second	Watts	1054.4
B.t.u. per square foot per hour	Joules per square meter per second	3.1546
B.t.u. per square foot per minute	Kilowatts per square foot	0.1758
B.t.u. per square foot per second for a temperature gradient of 1°F. per inch	Calories, gram (15°C.), per square centimeter per second for a temperature gradient of 1°C. per centimeter	1.2405
Drams (avoirdupois)	Grams	1.7719
Dynes	Newtons	1×10^{-5}
Ergs	Joules	1×10^{-7}
Faradays	Coulombs (abs.)	96,500
Fathoms	Feet	6
Feet	Meters	0.3048
Feet per minute	Centimeters per second	0.5080
Feet per minute	Miles per hour	0.011364
Feet per $(second)^2$	Meters per $(second)^2$	0.3048
Feet of water at 39.2°F.	Newtons per square meter	2989
Foot-poundals	B.t.u.	3.995×10^{-5}
Foot-poundals	Joules	0.04214
Foot-poundals	Liter-atmospheres	4.159×10^{-4}
Foot-pounds	B.t.u.	0.0012856
Foot-pounds	Calories, gram	0.3239
Foot-pounds	Foot-poundals	32.174
Foot-pounds	Horsepower-hours	5.051×10^{-7}
Foot-pounds	Kilowatt-hours	3.766×10^{-7}
Foot-pounds	Liter-atmospheres	0.013381
Foot-pounds force	Joules	1.3558
Foot-pounds per second	Horsepower	0.0018182
Foot-pounds per second	Kilowatts	0.0013558
Furlongs	Miles	0.125
Gallons (U.S. liquid)	Barrels (U.S. liquid)	0.03175
Gallons	Cubic meters	0.003785
Gallons	Cubic feet	0.13368
Gallons	Gallons (Imperial)	0.8327
Gallons	Liters	3.785
Gallons	Ounces (U.S. fluid)	128
Gallons per minute	Cubic feet per hour	8.021
Gallons per minute	Cubic feet per second	0.002228
Grains	Grams	0.06480
Grains	Pounds	1/7000
Grains per cubic foot	Grams per cubic meter	2.2884
Grains per gallon	Parts per million	17.118
Grams	Drams (avoirdupois)	0.5644
Grams	Drams (troy)	0.2572
Grams	Grains	15.432
Grams	Kilograms	0.001
Grams	Pounds (avoirdupois)	0.0022046
Grams	Pounds (troy)	0.002679
Grams per cubic centimeter	Pounds per cubic foot	62.43
Grams per cubic centimeter	Pounds per gallon	8.345
Grams per liter	Grains per gallon	58.42
Grams per liter	Pounds per cubic foot	0.0624
Grams per square centimeter	Pounds per square foot	2.0482
Grams per square centimeter	Pounds per square inch	0.014223
Hectares	Acres	2.471
Hectares	Square meters	10,000
Horsepower (British)	B.t.u. per minute	42.42
Horsepower (British)	B.t.u. per hour	2545
Horsepower (British)	Foot-pounds per minute	33,000
Horsepower (British)	Foot-pounds per second	550
Horsepower (British)	Watts	745.7

B.t.u. (60°F.) per degree Fahrenheit	Calories per degree centigrade	453.6
Bushels (U.S. dry)	Cubic feet	1.2444
Bushels (U.S. dry)	Cubic meters	0.03524
Calories, gram	B.t.u.	3.968×10^{-3}
Calories, gram	Foot-pounds	3.087
Calories, gram	Joules	4.1868
Calories, gram	Liter-atmospheres	4.130×10^{-2}
Calories, gram	Horsepower-hours	1.5591×10^{-6}
Calories, gram, per gram per degree C.	Joules per kilogram per degree Kelvin	4186.8
Calories, kilogram	Kilowatt-hours	0.0011626
Calories, kilogram per second	Kilowatts	4.185
Candle power (spherical)	Lumens	12.556
Carats (metric)	Grams	0.2
Centigrade heat units	B.t.u.	1.8
Centimeters	Angstrom units	1×10^{8}
Centimeters	Feet	0.03281
Centimeters	Inches	0.3937
Centimeters	Meters	0.01
Centimeters	Microns	10,000
Centimeters of mercury at 0°C.	Atmospheres	0.013158
Centimeters of mercury at 0°C.	Feet of water at 39.1°F.	0.4460
Centimeters of mercury at 0°C	Newtons per square meter	1333.2
Centimeters of mercury at 0°C.	Pounds per square foot	27.845
Centimeters of mercury at 0°C.	Pounds per square inch	0.19337
Centimeters per second	Feet per minute	1.9685
Centimeters of water at 4°C.	Newtons per square meter	98.064
Centistokes	Square meters per second	1×10^{-6}
Circular mils	Square centimeters	5.067×10^{-6}
Circular mils	Square inches	7.854×10^{-7}
Circular mils	Square mils	0.7854
Cords	Cubic feet	128
Cubic centimeters	Cubic feet	3.532×10^{-5}
Cubic centimeters	Gallons	2.6417×10^{-4}
Cubic centimeters	Ounces (U.S. fluid)	0.03381
Cubic centimeters	Quarts (U.S. fluid)	0.0010567
Cubic feet	Bushels (U.S.)	0.8036
Cubic feet	Cubic centimeters	28,317
Cubic feet	Cubic meters	0.028317
Cubic feet	Cubic yards	0.03704
Cubic feet	Gallons	7.481
Cubic feet	Liters	28.316
Cubic foot-atmospheres	Foot-pounds	2116.3
Cubic foot-atmospheres	Liter-atmospheres	28.316
Cubic feet of water (60°F.)	Pounds	62.37
Cubic feet per minute	Cubic centimeters per second	472.0
Cubic feet per minute	Gallons per second	0.1247
Cubic feet per second	Gallons per minute	448.8
Cubic feet per second	Million gallons per day	0.64632
Cubic inches	Cubic meters	1.6387×10^{-5}
Cubic yards	Cubic meters	0.76456
Curies	Disintegrations per minute	2.2×10^{12}
Curies	Coulombs per minute	1.1×10^{12}
Degrees	Radians	0.017453
Drams (apothecaries' or troy)	Grams	3.888

Horsepower (British)	Horsepower (metric)	1.0139
Horsepower (British)	Pounds carbon to CO_2 per hour	0.175
Horsepower (British)	Pounds water evaporated per hour at 212°F	2.64
Horsepower (metric)	Foot-pounds per second	542.47
Horsepower (metric)	Kilogram-meters per second	75.0
Hours (mean solar)	Seconds	3600
Inches	Meters	0.0254
Inches of mercury at 60°F	Newtons per square meter	3376.9
Inches of water at 60°F	Newtons per square meter	248.84
Joules (absolute)	B.t.u. (mean)	9.480×10^{-4}
Joules (absolute)	Calories, gram (mean)	0.2389
Joules (absolute)	Cubic foot-atmospheres	0.3485
Joules (absolute)	Foot-pounds	0.7376
Joules (absolute)	Kilowatt-hours	2.7778×10^{-7}
Joules (absolute)	Liter-atmospheres	0.009869
Kilocalories	Joules	4186.8
Kilograms	Pounds (avoirdupois)	2.2046
Kilograms force	Newtons	9.807
Kilograms per square centimeter	Pounds per square inch	14.223
Kilometers	Miles	0.6214
Kilowatt-hours	B.t.u.	3414
Kilowatt-hours	Foot-pounds	2.6552×10^{6}
Kilowatts	Horsepower	1.3410
Knots (international)	Meters per second	0.5144
Knots (nautical miles per hour)	Miles per hour	1.1516
Lamberts	Candles per square inch	2.054
Liter-atmospheres	Cubic foot-atmospheres	0.03532
Liter-atmospheres	Foot-pounds	74.74
Liters	Cubic feet	0.03532
Liters	Cubic meters	0.001
Liters	Gallons	0.26418
Lumens	Watts	0.001496
Micromicrons	Microns	1×10^{-6}
Microns	Angstrom units	1×10^{4}
Microns	Meters	1×10^{-6}
Miles (nautical)	Feet	6080
Miles (nautical)	Miles (U.S. statute)	1.1516
Miles	Feet	5280
Miles	Meters	1609.3
Miles per hour	Feet per second	1.4667
Miles per hour	Meters per second	0.4470
Milliliters	Cubic centimeters	1
Millimeters	Meters	0.001
Millimeters of mercury at 0°C.	Newtons per square meter	133.32
Millimicrons	Microns	0.001
Mils	Inches	0.001
Mils	Meters	2.54×10^{-5}
Minims (U.S.)	Cubic centimeters	0.06161
Minutes (angle)	Radians	2.909×10^{-4}
Minutes (mean solar)	Seconds	60
Newtons	Kilograms	0.10197
Ounces (avoirdupois)	Kilograms	0.02835
Ounces (avoirdupois)	Ounces (troy)	0.9115
Ounces (U.S. fluid)	Cubic meters	2.957×10^{-5}
Ounces (troy)	Ounces (apothecaries')	1.000

TABLE 1-6 Alphabetical Listing of Common Conversions *(Concluded)*

To convert from	To	Multiply by
Pints (U.S. liquid)	Cubic meters	4.732×10^{-4}
Poundals	Newtons	0.13826
Pounds (avoirdupois)	Grains	7000
Pounds (avoirdupois)	Kilograms	0.45359
Pounds (avoirdupois)	Pounds (troy)	1.2153
Pounds per cubic foot	Grams per cubic centimeter	0.016018
Pounds per cubic foot	Kilograms per cubic meter	16.018
Pounds per square foot	Atmospheres	4.725×10^{-4}
Pounds per square foot	Kilograms per square meter	4.882
Pounds per square inch	Atmospheres	0.06805
Pounds per square inch	Kilograms per square centimeter	0.07031
Pounds per square inch	Newtons per square meter	6894.8
Pounds force	Newtons	4.4482
Pounds force per square foot	Newtons per square meter	47.88
Pounds water evaporated from and at 212°F.	Horsepower-hours	0.379
Pound-centigrade units (p.c.u.)	B.t.u.	1.8
Quarts (U.S. liquid)	Cubic meters	9.464×10^{-4}
Radians	Degrees	57.30
Revolutions per minute	Radians per second	0.10472
Seconds (angle)	Radians	4.848×10^{-6}
Slugs	Gee pounds	1
Slugs	Kilograms	14.594
Slugs	Pounds	32.17

To convert from	To	Multiply by
Square centimeters	Square feet	0.0010764
Square feet	Square meters	0.0929
Square feet per hour	Square meters per second	2.581×10^{-5}
Square inches	Square centimeters	6.452
Square inches	Square meters	6.452×10^{-4}
Square yards	Square meters	0.8361
Stokes	Square meters per second	1×10^{-4}
Tons (long)	Kilograms	1016
Tons (long)	Pounds	2240
Tons (metric)	Kilograms	1000
Tons (metric)	Pounds	2204.6
Tons (metric)	Tons (short)	1.1023
Tons (short)	Kilograms	907.18
Tons (short)	Pounds	2000
Tons (refrigeration)	B.t.u. per hour	12,000
Tons (British shipping)	Cubic feet	42.00
Tons (U.S. shipping)	Cubic feet	40.00
Torr (mm. mercury, 0°C.)	Newtons per square meter	133.32
Watts	B.t.u. per hour	3.413
Watts	Joules per second	1
Watts	Kilogram-meters per second	0.10197
Watt-hours	Joules	3600
Yards	Meters	0.9144

TABLE 1-7 Common Units and Conversion Factors*

Mass (M)
1 pound mass = 453.5924 grams
= 0.45359 kilograms
= 7000 grains
1 slug = 32.174 pounds mass
1 ton (short) = 2000 pounds mass
1 ton (long) = 2240 pounds mass
1 ton (metric) = 1000 kilograms
= 2204.62 pounds mass
1 pound mole = 453.59 gram moles

Length (L)
1 foot = 30.480 centimeters
= 0.3048 meters
1 inch = 2.54 centimeters
= 0.0254 meters
1 mile (U.S.) = 1.60935 kilometers
1 yard = 0.9144 meters

Area (L^2)
1 square foot = 929.0304 square centimeters
= 0.09290304 square meters
1 square inch = 6.4516 square centimeters
1 square yard = 0.836127 square meters

Volume (L^3)
1 cubic foot = 28,316.85 cubic centimeters
= 0.02831685 cubic meters
= 28.31685 liters
= 7.481 gallons (U.S.)
1 gallon = 3.7853 liters
= 231 cubic inches

Time (θ)
1 hour = 60 minutes
= 3600 seconds

Temperature (T)
1 centigrade or Celsius degree = 1.8 Fahrenheit degree
Temperature, Kelvin = T°C + 273.15
Temperature, Rankine = T°F + 459.7
Temperature, Fahrenheit = 9/5 T°C + 32
Temperature, centigrade or Celsius = 5/9 (T°F − 32)
Temperature, Rankine = 1.8 T K

Force (F)
1 pound force = 444,822.2 dynes
= 4.448222 Newtons
= 32.174 poundals

Pressure (F/L^2)
Normal atmospheric pressure
1 atm = 760 millimeters of mercury at 0°C (density 13.5951 g/cm^3)
= 29.921 inches of mercury at 32°F
= 14.696 pounds force/square inch
= 33.899 feet of water at 39.1°F
= 1.01325×10^6 dynes/square centimeter
= 1.01325×10^5 Newtons/square meter

Density (M/L^3)
1 pound mass/cubic foot = 0.01601846 grams/cubic centimeter
= 16.01846 kilogram/cubic meter

Energy (H or FL)
1 British thermal unit = 251.98 calories
= 1054.4 joules
= 777.97 foot-pounds force
= 10.409 liter-atmospheres
= 0.2930 watt-hour

Diffusivity (L^2/θ)
1 square foot/hour = 0.258 cm^2/s
= 2.58×10^{-5} m^2/s

Viscosity ($M/L\theta$)
1 pound mass/foot hour = 0.00413 g/cm s
0.000413 kg/m s
1 centipoise = 0.01 poise
= 0.01 g/cm s
= 0.001 kg/m s
= 0.000672 lbm/ft s
= 0.0000209 $lb_f s/ft^2$

Thermal conductivity [$H/\theta L^2(T/L)$]
1 Btu/hr ft^2 (°F/ft) = 0.00413 cal/s cm^2 (°C/cm)
= 1.728 J/s m^2 (°C/m)

Heat transfer coefficient
1 Btu/hr ft^2 °F = 5.678 J/s m^2 °C

Heat capacity (H/MT)
1 Btu/lbm °F = 1 cal/g °C
= 4184 J/kg °C

Gas constant
1.987 Btu/lbm mole °R = 1.987 cal/mol K
= 82.057 atm cm^3/mol K
= 0.7302 atm ft^3/lb mole °F
= 10.73 (lb_f/in.^{2}) (ft^3)/lb mole °R
= 1545 (lb_f/ft^2) (ft^3)/lb mole °R
= 8.314 (N/m^2) (m^3)/mol K

Gravitational acceleration
g = 9.8066 m/s^2
= 32.174 ft/s^2

NOTE: U.S. customary units, or British units, on left and SI units on right.
*Adapted from Faust et al., *Principles of Unit Operations,* John Wiley and Sons, 1980.

TABLE 1-8 Kinematic-Viscosity Conversion Formulas

Viscosity scale	Range of t, sec	Kinematic viscosity, stokes
Saybolt Universal	$32 < t < 100$	$0.00226t - 1.95/t$
	$t > 100$	$0.00220t - 1.35/t$
Saybolt Furol	$25 < t < 40$	$0.0224t - 1.84/t$
	$t > 40$	$0.0216t - 0.60/t$
Redwood No. 1	$34 < t < 100$	$0.00260t - 1.79/t$
	$t > 100$	$0.00247t - 0.50/t$
Redwood Admiralty		$0.027t - 20/t$
Engler		$0.00147t - 3.74/t$

TABLE 1-9 Values of the Gas-Law Constant

Temp. scale	Press. units	Vol. units	Wt. units	Energy units*	R
Kelvin			g-moles	calories	1.9872
			g-moles	joules (abs)	8.3144
			g-moles	joules (int)	8.3130
	atm.	cm^3	g-moles	atm cm^3	82.057
	atm.	liters	g-moles	atm liters	0.08205
	mm. Hg	liters	g-moles	mm Hg-liters	62.361
	bar	liters	g-moles	bar-liters	0.08314
	kg/cm^2	liters	g-moles	kg/(cm^2)(liters)	0.08478
	atm	ft^3	lb-moles	atm-ft^3	1.314
	mm Hg	ft^3	lb-moles	mm Hg-ft^3	998.9
			lb-moles	chu or pcu	1.9872
Rankine			lb-moles	Btu	1.9872
			lb-moles	hp-hr	0.0007805
			lb-moles	kw-hr	0.0005819
	atm	ft^3	lb-moles	atm-ft^3	0.7302
	in Hg	ft^3	lb-moles	in Hg-ft^3	21.85
	mm Hg	ft^3	lb-moles	mm Hg-ft^3	555.0
	lb/in^{2}abs	ft^3	lb-moles	(lb)(ft^3)/in^2	10.73
	lb/ft^{2}abs	ft^3	lb-moles	ft-lb	1545.0

*Energy units are the product of pressure units and volume units.

TABLE 1-10 United States Customary System of Weights and Measures

Linear Measure

12 inches (in) or (″) = 1 foot (ft) or (′)
3 feet = 1 yard (yd)
16.5 feet, 5.5 yards = 1 rod (rd)
5280 feet, 320 rods = 1 mile (mi)
1 mil = 0.001 inch

Nautical:

6080.2 feet = 1 nautical mile
6 feet = 1 fathom
120 fathoms = 1 cable length
1 knot = 1 nautical mile per hour
60 nautical miles = 1° of latitude

Square Measure

144 sq. inches (sq. in) or (in^2) or (□″) = 1 sq. foot (ft^2) or (□′)
9 sq. feet (ft^2) (□′) = 1 sq. yard (yd^2)
30.25 sq. yards = 1 sq. rod, pole, or perch
160 sq. rods = {10 sq. chains, 43,560 sq. ft} = 1 acre
640 acres = 1 sq. mile = 1 section
1 circular inch (area of circle of 1 inch diameter) = 0.7854 sq. inch
1 sq. inch = 1.2732 circular inch
1 circular mil = area of circle of 0.001 inch diameter
1,000,000 circular mils = 1 circular inch

Circular Measure

60 seconds (″) (sec) = 1 minute (min) or (′)
60 minutes (′) = 1 degree (°)
90 degrees (°) = 1 quadrant
360 degrees (°) = 1 circumference
57.29578 degrees {= 1 radian (rad.), = 57° 17′ 44.81″}

Volume Measure

Solid:

1728 cubic in (cu. in) (in^3) = 1 cubic foot (cu. ft)(ft^3)
27 cu. ft = 1 cubic yard (cu. yd)

Dry Measure:

2 pints = 1 quart
8 quarts = 1 peck
4 pecks = 1 bushel
1 United States Winchester bushel = 2150.42 cubic inches

Liquid:

4 gills = 1 pint (pt)
2 pints = 1 quart (qt)
4 quarts = 1 gallon (gal)
7.4805 gallons = 1 cubic foot

Apothecaries' Liquid:

60 minims (min. or ♏) = 1 fluid dram or drachm
8 drams (ʒ) = 1 fluid ounce
16 ounces (oz. ℥) = 1 pint

Avoirdupois Weight

16 drams = 437.5 grains = 1 ounce (oz)
16 ounces = 7000 grains = 1 pound (lb)
100 pounds = 1 hundredweight (cwt)
2000 pounds = 1 short ton: 2240 pounds = 1 long ton

Troy Weight

24 grains = 1 pennyweight (dwt)
20 pennyweights = 1 ounce (oz)
12 ounces = 1 pound (lb)

Apothecaries' Weight

20 grains (gr) = 1 scruple (℈)
3 scruples = 1 dram (ʒ)
8 drams = 1 ounce (℥)
12 ounces = 1 pound (lb)

TABLE 1-11 Temperature Conversion Formulas

°F = (°C × 5/9) + 32
°C = (°F − 32) × 5/9
°R = °F + 459.69
°K = °C + 273.15
°K = °R × 5/9

Temperature difference, ΔT
°F = °C × 9/5

NOTE: An extensive table of temperature conversions may be found in the sixth edition of the *Handbook* (Table 1-12).

TABLE 1-12 Greek Alphabet

Alpha	= A, α = A, a	Nu	= N, ν = N, n
Beta	= B, β = B, b	Xi	= Ξ, ξ = X, x
Gamma	= Γ, γ = G, g	Omicron	= O, o = O, o
Delta	= Δ, δ = D, d	Pi	= Π, π = P, p
Epsilon	= E, ε = E, e	Rho	= P, ρ = R, r
Zeta	= Z, ζ = Z, z	Sigma	= Σ, σ = S, s
Eta	= H, η = E, e	Tau	= T, τ = T, t
Theta	= Θ, θ = Th, th	Upsilon	= Y, υ = U, u
Iota	= I, ι = I, i	Phi	= Φ, ϕ = Ph, ph
Kappa	= K, κ = K, k	Chi	= X, χ = Ch, ch
Lambda	= Λ, λ = L, l	Psi	= Ψ, ψ = Ps, ps
Mu	= M, μ = M, m	Omega	= Ω, ω = O, o

TABLE 1-13 Specific Gravity, Degrees Baumé, Degrees API, Degrees Twaddell, Pounds per Gallon, Pounds per Cubic Foot*

$$°Bé = 145 - \frac{145}{\text{sp gr}} \text{ (heavier than } H_2O\text{)}; \quad °Bé = \frac{140}{\text{sp gr}} - 130 \text{ (lighter than } H_2O\text{)}; \quad °Tw = \frac{\text{sp gr } 60°/60°F - 1}{0.005} \quad °API = \frac{141.5}{\text{sp gr}} - 131.5$$

Sp gr 60°/60°	°Bé	°API	Lb per gal at 60°F wt in air	Lb per ft³ at 60°F wt in air	Sp gr 60°/60°	°Bé	°API	Lb per gal at 60°F wt in air	Lb per ft³ at 60°F wt in air	Sp gr 60°/60°	°Bé	°API	Lb per gal at 60°F wt in air	Lb per ft³ at 60°F wt in air	Sp gr 60°/60°	°Bé	°API	Lb per gal at 60°F wt in air	Lb per ft³ at 60°F. wt. in air
0.600	103.33	104.33	4.9929	37.350	0.700	70.00	70.64	5.8268	43.587	0.800	45.00	45.38	6.6606	49.825	0.900	25.56	25.72	7.4944	56.062
.605	101.40	102.38	5.0346	37.662	.705	68.58	69.21	5.8685	43.899	.805	43.91	44.28	6.7023	50.137	.905	24.70	24.85	7.5361	56.374
.610	99.51	100.47	5.0763	37.973	.710	67.18	67.80	5.9101	44.211	.810	42.84	43.19	6.7440	50.448	.910	23.85	23.99	7.5777	56.685
.615	97.64	98.58	5.1180	38.285	.715	65.80	66.40	5.9518	44.523	.815	41.78	42.12	6.7857	50.760	.915	23.01	23.14	7.6194	56.997
.620	95.81	96.73	5.1597	38.597	.720	64.44	65.03	5.9935	44.834	.820	40.73	41.06	6.8274	51.072	.920	22.17	22.30	7.6612	57.310
.625	94.00	94.90	5.2014	39.910	.725	63.10	63.67	6.0352	45.146	.825	39.70	40.02	6.8691	51.384	.925	21.35	21.47	7.7029	57.622
.630	92.22	93.10	5.2431	39.222	.730	61.78	62.34	6.0769	45.458	.830	38.67	38.98	6.9108	51.696	.930	20.54	20.65	7.7446	57.934
.635	90.47	91.33	5.2848	39.534	.735	60.48	61.02	6.1186	45.770	.835	37.66	37.96	6.9525	52.008	.935	19.73	19.84	7.7863	58.246
.640	88.75	89.59	5.3265	39.845	.740	59.19	59.72	6.1603	46.082	.840	36.67	36.95	6.9941	52.320	.940	18.94	19.03	7.8280	58.557
.645	87.05	87.88	5.3682	40.157	.745	57.92	58.43	6.2020	46.394	.845	35.68	35.96	7.0358	52.632	.945	18.15	18.24	7.8697	58.869
.650	85.38	86.19	5.4098	40.468	.750	56.67	57.17	6.2437	46.706	.850	34.71	34.97	7.0775	52.943	.950	17.37	17.45	7.9114	59.181
.655	83.74	84.53	5.4515	40.780	.755	55.43	55.92	6.2854	47.018	.855	33.74	34.00	7.1192	53.255	.955	16.60	16.67	7.9531	59.493
.660	82.12	82.89	5.4932	41.092	.760	54.21	54.68	6.3271	47.330	.860	32.79	33.03	7.1609	53.567	.960	15.83	15.90	7.9947	59.805
.665	80.53	81.28	5.5349	41.404	.765	53.01	53.47	6.3688	47.642	.865	31.85	32.08	7.2026	53.879	.965	15.08	15.13	8.0364	60.117
.670	78.96	79.69	5.5766	41.716	.770	51.82	52.27	6.4104	47.953	.870	30.92	31.14	7.2443	54.191	.970	14.33	14.38	8.0780	60.428
.675	77.41	78.13	5.6183	42.028	.775	50.65	51.08	6.4521	47.265	.875	30.00	30.21	7.2860	54.503	.975	13.59	13.63	8.1197	60.740
.680	75.88	76.59	5.6600	42.340	.780	49.49	49.91	6.4938	48.577	.880	29.09	29.30	7.3277	54.815	.980	12.86	12.89	8.1615	61.052
.685	74.38	75.07	5.7017	42.652	.785	48.34	48.75	6.5355	48.889	.885	28.19	28.39	7.3694	55.127	.985	12.13	12.15	8.2032	61.364
.690	72.90	73.57	5.7434	42.963	.790	47.22	47.61	6.5772	49.201	.890	27.30	27.49	7.4111	55.438	.990	11.41	11.43	8.2449	61.676
.695	71.44	72.10	5.7851	43.275	.795	46.10	46.49	6.6189	49.513	.895	26.42	26.60	7.4528	55.750	.995	10.70	10.71	8.2866	61.988
															1.000	10.00	10.00	8.3283	62.300

Sp gr 60°/60°	°Bé	°Tw	Lb per gal at 60°F wt in air	Lb per ft³ at 60°F wt in air	Sp gr 60°/60°	°Bé	°Tw	Lb per gal at 60°F wt in air	Lb per ft³ at 60°F wt in air	Sp gr 60°/60°	°Bé	°Tw	Lb per gal at 60°F wt in air	Lb per ft³ at 60°F wt in air	Sp gr 60°/60°	°Bé	°Tw	Lb per gal at 60°F wt in air	Lb per ft³ at 60°F. wt. in air
1.005	0.72	1	8.3700	62.612	1.255	29.46	51	10.4546	78.206	1.505	48.65	101	12.5392	93.800	1.755	62.38	151	14.6238	109.394
1.010	1.44	2	8.4117	62.924	1.260	29.92	52	10.4963	78.518	1.510	48.97	102	12.5809	94.112	1.760	62.61	152	14.6655	109.705
1.015	2.14	3	8.4534	63.236	1.265	30.38	53	10.5380	78.830	1.515	49.29	103	12.6226	94.424	1.765	62.85	153	14.7072	110.017
1.020	2.84	4	8.4950	63.547	1.270	30.83	54	10.5797	79.141	1.520	49.61	104	12.6643	94.735	1.770	63.08	154	14.7489	110.329
1.025	3.54	5	8.5367	63.859	1.275	31.27	55	10.6214	79.453	1.525	49.92	105	12.7060	95.047	1.775	63.31	155	14.7906	110.641
1.030	4.22	6	8.5784	64.171	1.280	31.72	56	10.6630	79.765	1.530	50.23	106	12.7477	95.359	1.780	63.54	156	14.8323	110.953
1.035	4.90	7	8.6201	64.483	1.285	32.16	57	10.7047	80.077	1.535	50.54	107	12.7894	95.671	1.785	63.77	157	14.8740	111.265
1.040	5.58	8	8.6618	64.795	1.290	32.60	58	10.7464	80.389	1.540	50.84	108	12.8310	95.983	1.790	63.99	158	14.9157	111.577
1.045	6.24	9	8.7035	65.107	1.295	33.03	59	10.7881	80.701	1.545	51.15	109	12.8727	96.295	1.795	64.22	159	14.9574	111.889
1.050	6.91	10	8.7452	65.419	1.300	33.46	60	10.8298	81.013	1.550	51.45	110	12.9144	96.606	1.800	64.44	160	14.9990	112.200
1.055	7.56	11	8.7869	65.731	1.305	33.89	61	10.8715	81.325	1.555	51.75	111	12.9561	96.918	1.805	64.67	161	15.0407	112.512
1.060	8.21	12	8.8286	66.042	1.310	34.31	62	10.9132	81.636	1.560	52.05	112	12.9978	97.230	1.810	64.89	162	15.0824	112.824
1.065	8.85	13	8.8703	66.354	1.315	34.73	63	10.9549	81.948	1.565	52.35	113	13.0395	97.542	1.815	65.11	163	15.1241	113.136
1.070	9.49	14	8.9120	66.666	1.320	35.15	64	10.9966	82.260	1.570	52.64	114	13.0812	97.854	1.820	65.33	164	15.1658	113.448
1.075	10.12	15	8.9537	66.978	1.325	35.57	65	11.0383	82.572	1.575	52.94	115	13.1229	98.166	1.825	65.55	165	15.2075	113.760
1.080	10.74	16	8.9954	67.290	1.330	35.98	66	11.0800	82.884	1.580	53.23	116	13.1646	98.478	1.830	65.77	166	15.2492	114.072
1.085	11.36	17	9.0371	67.602	1.335	36.39	67	11.1217	83.196	1.585	53.52	117	13.2063	98.790	1.835	65.98	167	15.2909	114.384
1.090	11.97	18	9.0787	67.914	1.340	36.79	68	11.1634	83.508	1.590	53.81	118	13.2480	99.102	1.840	66.20	168	15.3326	114.696
1.095	12.58	19	9.1204	68.226	1.345	37.19	69	11.2051	83.820	1.595	54.09	119	13.2897	99.414	1.845	66.41	169	15.3743	115.007
1.100	13.18	20	9.1621	68.537	1.350	37.59	70	11.2467	84.131	1.600	54.38	120	13.3313	99.725	1.850	66.62	170	15.4160	115.318
1.105	13.78	21	9.2038	68.849	1.355	37.99	71	11.2884	84.443	1.605	54.66	121	13.3730	100.037	1.855	66.83	171	15.4577	115.630
1.110	14.07	22	9.2455	69.161	1.360	38.38	72	11.3301	84.755	1.610	54.94	122	13.4147	100.349	1.860	67.04	172	15.4993	115.943
1.115	14.96	23	9.2872	69.473	1.365	38.77	73	11.3718	85.067	1.615	55.22	123	13.4564	100.661	1.865	67.25	173	15.5410	116.255
1.120	15.54	24	9.3289	69.785	1.370	39.16	74	11.4135	85.379	1.620	55.49	124	13.4981	100.973	1.870	67.46	174	15.5827	116.567
1.125	16.11	25	9.3706	70.097	1.375	39.55	75	11.4552	85.691	1.625	55.77	125	13.5398	101.285	1.875	67.67	175	15.6244	116.879
1.130	16.68	26	9.4123	70.409	1.380	39.93	76	11.4969	86.003	1.630	56.04	126	13.5815	101.597	1.880	67.87	176	15.6661	117.191
1.135	17.25	27	9.4540	70.721	1.385	40.31	77	11.5386	86.315	1.635	56.32	127	13.6232	101.909	1.885	68.08	177	15.7078	117.503
1.140	17.81	28	9.4957	71.032	1.390	40.68	78	11.5803	86.626	1.640	56.59	128	13.6649	102.220	1.890	68.28	178	15.7495	117.814
1.145	18.36	29	9.5374	71.344	1.395	41.06	79	11.6220	86.938	1.645	56.85	129	13.7066	102.532	1.895	68.48	179	15.7912	118.126
1.150	18.91	30	9.5790	71.656	1.400	41.43	80	11.6637	87.250	1.650	57.12	130	13.7483	102.844	1.900	68.68	180	15.8329	118.438
1.155	19.46	31	9.6207	71.968	1.405	41.80	81	11.7054	87.562	1.655	57.39	131	13.7900	103.156	1.905	68.88	181	15.8746	118.740
1.160	20.00	32	9.6624	72.280	1.410	42.16	82	11.7471	87.874	1.660	57.65	132	13.8317	103.468	1.910	69.08	182	15.9163	119.062
1.165	20.54	33	9.7041	72.592	1.415	42.53	83	11.7888	88.186	1.665	57.91	133	13.8734	103.780	1.915	69.28	183	15.9580	119.374
1.170	21.07	34	9.7458	72.904	1.420	42.89	84	11.8304	88.498	1.670	58.17	134	13.9150	104.092	1.920	69.48	184	15.9996	119.686
1.175	21.60	35	9.7875	73.216	1.425	43.25	85	11.8721	88.810	1.675	58.43	135	13.9567	104.404	1.925	69.68	185	16.0413	119.998
1.180	22.12	36	9.8292	73.528	1.430	43.60	86	11.9138	89.121	1.680	58.69	136	13.9984	104.715	1.930	69.87	186	16.0830	120.309
1.185	22.64	37	9.8709	73.840	1.435	43.95	87	11.9555	89.433	1.685	58.95	137	14.0401	105.027	1.935	70.06	187	16.1247	120.621
1.190	23.15	38	9.9126	74.151	1.440	44.31	88	11.9972	89.745	1.690	59.20	138	14.0818	105.339	1.940	70.26	188	16.1664	120.933
1.195	23.66	39	9.9543	74.463	1.445	44.65	89	12.0389	90.057	1.695	59.45	139	14.1235	105.651	1.945	70.45	189	16.2081	121.245
1.200	24.17	40	9.9960	74.775	1.450	45.00	90	12.0806	90.369	1.700	59.71	140	14.1652	105.963	1.950	70.64	190	16.2498	121.557
1.205	24.67	41	10.0377	75.087	1.455	45.34	91	12.1223	90.681	1.705	59.96	141	14.2069	106.275	1.955	70.83	191	16.2915	121.869
1.210	25.17	42	10.0793	75.399	1.460	45.68	92	12.1640	90.993	1.710	60.20	142	14.2486	106.587	1.960	71.02	192	16.3332	122.181
1.215	25.66	43	10.1210	75.711	1.465	46.02	93	12.2057	91.305	1.715	60.45	143	14.2903	106.899	1.965	71.21	193	16.3749	122.493
1.220	26.15	44	10.1627	76.022	1.470	46.36	94	12.2473	91.616	1.720	60.70	144	14.3320	107.210	1.970	71.40	194	16.4166	122.804
1.225	26.63	45	10.2044	76.334	1.475	46.69	95	12.2890	91.928	1.725	60.94	145	14.3737	107.522	1.975	71.58	195	16.4583	123.116
1.230	27.11	46	10.2461	76.646	1.480	47.03	96	12.3307	92.240	1.730	61.18	146	14.4153	107.834	1.980	71.77	196	16.5000	123.428
1.235	27.59	47	10.2878	76.958	1.485	47.36	97	12.3724	92.552	1.735	61.34	147	14.4570	108.146	1.985	71.95	197	16.5417	123.740
1.240	28.06	48	10.3295	77.270	1.490	47.68	98	12.4141	92.864	1.740	61.67	148	14.4987	108.458	1.990	72.14	198	16.5833	124.052
1.245	28.53	49	10.3712	77.582	1.495	48.01	99	12.4558	93.176	1.745	61.91	149	14.5404	108.770	1.995	72.32	199	16.6250	124.364
1.250	29.00	50	10.4129	77.894	1.500	48.33	100	12.4975	93.488	1.750	62.14	150	14.5821	109.082	2.000	72.50	200	16.6667	124.676

*Prepared by Lewis V. Judson, Ph.D., Chief of Length Section of National Bureau of Standards with the advice and assistance of E. L. Peffer, B.S., A.M., late Chief of Capacity and Density Section, National Bureau of Standards.

TABLE 1-14 Fundamental Physical Constants

1 sec = 1.00273791 sidereal seconds	sec = mean solar second
g_0 = 9.80665 m/sec^2	Definition: g_0 = standard gravity
1 liter = 0.001 cu. m	
1 atm = 101,325 newtons/sq m	Definition: atm = standard atmosphere
1 mm Hg (pressure) = ($1/760$) atm	mm Hg (pressure) = standard millimeter mercury
= 133.3224 newtons/sq m	
1 int ohm = 1.000495 ± 0.000015 abs ohm	int = international; abs = absolute
1 int amp = 0.999835 ± 0.000025 abs amp	amp = ampere
1 int coul = 0.999835 ± 0.000025 abs coul	coul = coulomb
1 int volt = 1.000330 ± 0.000029 abs volt	
1 int watt = 1.000165 ± 0.000052 abs watt	
1 int joule = 1.000165 ± 0.000052 abs joule	
$T_{0°C}$ = 273.150 ± 0.010°K	Absolute temperature of the ice point, 0°C
$(PV)_{0°C}^{P=0} = (RT)_{0°C}$ = 2271.16 ± 0.04 abs joule/mole	
= 22,414.6 ± 0.4 cu. cm atm/mole	*PV* product for ideal gas at 0°C
= 22.4146 ± 0.0004 liter atm/mole	
R = 8.31439 ± 0.00034 abs joule/deg mole	*R* = gas constant per mole
= 1.98719 ± 0.00013 cal/deg mole	
= 82.0567 ± 0.0034 cu. cm atm/deg mole	
= 0.0820567 ± 0.0000034 liter atm/deg mole	
ln 10 = 2.302585	ln = natural logarithm (base e)
R ln 10 = 19.14460 ± 0.00078 abs joule/deg mole	
= 4.57567 ± 0.00030 cal/deg mole	
N = (6.02283 ± 0.0022) × 10^{23}/mole	*N* = *Avogadro number*
h = (6.6242 ± 0.0044) × 10^{-34} joule sec	*h* = Planck constant
c = (2.99776 ± 0.00008) × 10^{8} m/sec	*c* = velocity of light
$(h^2/8\pi^2k)$ = (4.0258 ± 0.0037) × 10^{-39} g sq cm deg	Constant in rotational partition function of gases
$(h/8\pi^2c)$ = (2.7986 ± 0.0018) × 10^{-39} g cm	Constant relating wave number and moment of inertia
Z = *Nhc* = 11.9600 ± 0.0036 abs joule cm/mole	Z = constant relating wave number and energy per mole
= 2.85851 ± 0.0009 cal cm/mole	
(*Z*/*R*) = (*hc*/*k*) = c_2 = 1.43847 ± 0.00045 cm deg	c_2 = second radiation constant
$\mathcal{F}$ = 96,501.2 ± 10.0 int coul/g-equiv or int joule/int volt g-equiv	$\mathcal{F}$ = Faraday constant
= 96,485.3 ± 10.0 abs coul/g-equiv or abs joule/abs volt g-equiv	
= 23,068.1 ± 2.4 cal/int volt g-equiv	
= 23,060.5 ± 2.4 cal/abs volt g-equiv	
e = (1.60199 ± 0.00060) × 10^{-19} abs coul	*e* = electronic charge
= (1.60199 ± 0.00060) × 10^{-20} abs emu	
= (4.80239 ± 0.00180) × 10^{-10} abs esu	
1 int electron-volt/molecule = 96,501.2 ± 10 int joule/mole	
= 23,068.1 ± 2.4 cal/mole	
1 abs electron-volt/molecule = 96,485.3 ± 10. abs joule/mole	
= 23,060.5 ± 2.4 cal/mole	
1 int electron-volt = (1.60252 ± 0.00060) × 10^{-12} erg	
1 abs electron-volt = (1.60199 ± 0.00060) × 10^{-12} erg	
hc = (1.23916 ± 0.00032) × 10^{-4} int electron-volt cm	Constant relating wave number and energy per molecule
= (1.23957 ± 0.00032) × 10^{-4} abs electron-volt cm	
k = (8.61442 ± 0.00100) × 10^{-5} int electron-volt/deg	*k* = Boltzmann constant
= (8.61727 ± 0.00100) × 10^{-5} abs electron-volt/deg	
= (*R*/*N*) = (1.38048 ± 0.00050) × 10^{-23} joule/deg	
1 IT cal = ($1/860$) = 0.00116279 int watt-hr	Definition of IT cal: IT = International steam tables
= 4.18605 int joule	
= 4.18674 abs joule	
= 1.000654 cal	cal = thermochemical calorie
1 cal = 4.1840 abs joule	Definition: cal = thermochemical calorie
= 4.1833 int joule	
= 41.2929 ± 0.0020 cu. cm atm	
= 0.0412929 ± 0.0000020 liter atm	
1 IT cal/g = 1.8 Btu/lb	Definition of Btu: Btu = IT British Thermal Unit
1 Btu = 251.996 IT cal	
= 0.293018 int watt-hr	
= 1054.866 int joule	
= 1055.040 abs joule	
= 252.161 cal	cal = thermochemical calorie
1 horsepower = 550 ft-lb (wt)/sec	Definition of horsepower (mechanical): lb (wt) = weight of 1 lb
= 745.578 int watt	at standard gravity
= 745.70 abs watt	Definition of in: in = U.S. inch
1 in = (1/0.3937) = 2.54 cm	ft = U.S. foot (1 ft = 12 in)
1 ft = 0.304800610 m	
1 lb = 453.5924277 g	Definition; lb = avoirdupois pound
1 gal = 231 cu. in	Definition; gal = U.S. gallon
= 0.133680555 cu. ft	
= 3.785412 × 10^{-3} cu. m	
= 3.785412 liter	

CONVERSION OF VALUES FROM U.S. CUSTOMARY UNITS TO SI UNITS

American engineers are probably more familiar with the magnitude of physical entities in U.S. customary units than in SI units. Consequently, errors made in the conversion from one set of units to the other may go undetected. The following six examples will show how to convert the elements in six dimensionless groups. Proper conversions will result in the same numerical value for the dimensionless number. The dimensionless numbers used as examples are the Reynolds, Prandtl, Nusselt, Grashof, Schmidt, and Archimedes numbers.

Table 1-7 provides a number of useful conversion factors. To make a conversion of an element in U.S. customary units to SI units, one multiplies the value of the U.S. customary unit, found on the left side in the table, by the equivalent value on the right side. For example, to convert 10 British thermal units to joules, one multiplies 10 by 1054.4 to obtain 10544 joules.

In each example, the initial values of the factors are expressed in U.S. customary units, and the dimensionless value is calculated. Then the factors are converted to SI units, and the dimensionless value is recalculated. The two dimensionless values will be approximately the same. (Small variations occur due to the number of significant figures carried in the solution.)

Example 1. Calculation of a Reynolds Number

$$N_{Re} = \frac{DV\rho}{\mu}$$

U.S. customary units

$D = 3$ in. $= 3/12$ ft

$V = 6$ ft/s

$\rho = 0.08$ lbm/ft^3

$\mu = 0.015$ cp $= (0.015)(0.000672)$ lbm/ft·s

$$N_{Re} = \frac{(3/12)(6)(0.08)}{(0.015)(0.000672)} = 11{,}904$$

SI units

$D = (3)(0.0254)$ m

$V = (6)(0.3048)$ m/s

$\rho = (0.08)(16.018)$ kg/m^3

$\mu = (0.015)(0.001)$ kg/m·s

$$N_{Re} = \frac{(3 \times 0.0254)\ (6 \times 0.3048)\ (0.08 \times 16.018)}{(0.015)\ (0.001)} = 11{,}904$$

Example 2. Calculation of a Prandtl Number

$$N_{Pr} = \frac{C_p\mu}{k}$$

U.S. customary units

$\gamma_p = 0.47$ Btu/lbm °F

$\mu = 15$ centipoise $= (15)\ (0.000672)\ (3600)$ lbm/ft·hr

$k = 0.065$ Btu/hr·ft^2 (°F/ft)

$$N_{Pr} = \frac{(0.47)\ (15 \times 0.000672 \times 3600)}{0.065} = 262.4$$

SI units

$\gamma = (0.47)(4184)$ J/kg °C

$\mu = (15)(0.001)$ kg/m·s

$k = (0.065)(1.728)$ J/s·m^2 (°C/m)

$$N_{Pr} = \frac{(0.47)\ (4184)\ (15)\ (0.001)}{(0.065)\ (1.728)} = 262.6$$

(Difference due to rounding)

Example 3. Calculation of a Nusselt Number

$$N_{Nu} = \frac{hD}{k}$$

U.S. customary units

$h = 200$ Btu/hr·ft^2·°F

$D = 1.5$ in. $= 1.5/12$ ft

$k = 0.07$ Btu/hr·ft^2 (°F/ft)

$$N_{Nu} = \frac{(200)(1.5/12)}{0.07} = 357.1$$

SI units

$h = (200)(5.678)$ J/(s·m^2·°C)

$D = (1.5)(0.0254)$ m

$k = (0.07)(1.728)$ J/s·m^2 (°C/m)

$$N_{Nu} = \frac{(200)\ (5.678)\ (1.5)\ (0.0254)}{(0.07)\ (1.728)} = 357.7$$

(Difference due to rounding)

Example 4. Calculation of a Grashof Number

$$N_{Gr} = L^3\rho^2 g\beta(\Delta T)/\mu^2$$

U.S. Customary units

$L = 3$ ft

$\rho = 0.0725$ lbm/ft^3

$g = 32.174$ ft/s^2

$\beta = 0.00168$/°R

$\Delta T = 99$ °R

$\mu = 0.019$ centipoise $= 0.019 \times 0.000672$ lbm/ft·s $= 1.277 \times 10^{-5}$ lbm/ft·s

$$N_{Gr} = \frac{(3^3)\ (0.0725)^2(32.174)\ (0.00168)\ (99)}{(1.277 \times 10^{-5})^2} = 4.66 \times 10^9$$

SI units

$L = (3)(0.3048) = 0.9144$ m

$\rho = (0.0725)(16.018) = 1.1613$ kg/m^3

$g = 9.807$ m/s^2

$\beta = (0.00168)/(1.8) = 0.000933$/°K

$\Delta T = (99)(1.8) = 178.2$ °K

$\mu = (0.019)(0.001) = 1.9 \times 10^{-5}$ kg/m·s

$$N_{Gr} = \frac{(0.9144)^3(1.1613)^2(9.807)(0.000933)(178.2)}{(1.9 \times 10^{-5})^2} = 4.66 \times 10^9$$

Example 5. Calculation of a Schmidt Number

$$N_{Sc} = \frac{\mu}{\rho D}$$

U.S. customary units

$\mu = 0.02$ centipoise $= (0.02)(2.42)$ lbm/ft·hr

$\rho = 0.08$ lbm/ft^3

$D = 1.0$ ft^2/hr (diffusivity)

$$N_{Sc} = \frac{(0.02)\ (2.42)}{(0.08)(1.0)} = 0.605$$

SI units

$\mu = (0.02)(0.001)$ kg/m·s

$\rho = (0.08)(16.02)$ kg/m^2

$D = (1.0)(2.58 \times 10^{-5})$ m^2/s

$$N_{Sc} = \frac{(0.02)\ (0.001)}{(0.08)(16.02)(1.0)\ (2.58 \times 10^{-5})} = 0.605$$

Example 6. Calculation of an Archimedes Number

$$N_{Ar} = \frac{d^3\rho_f(\rho_p - \rho_f)g}{\mu^2}$$

U.S. customary units

$d = 2$ mm $= 2/[(1000)(0.3048)] = 0.00656$ ft

$\rho_f = 0.0175$ lbm/ft^3

$\rho_p = 168.5$ lbm/ft^3

$g = 32.174$ ft/s^2

$\mu = 0.04$ centipoise $= 0.04 \times 0.000672 = 2.688^{-5}$ lbm/ft·s

$$N_{Ar} = \frac{(0.00656)^3\ (0.0175)\ (168.5 - 0.017)\ (32.174)}{(2.688 \times 10^{-5})^2} = 37{,}064$$

SI units

$d = 2/1000$ m

$\rho_p = 168.5 \times 16.02 = 2699.37$ kg/m^3

$\rho_f = 0.0175 \times 16.02 = 0.2804$ g/m^3

$g = 9.807$ m/s^2

$\mu = 0.04 \times 0.001 = 4 \times 10^{-5}$ kg/m·s

$$N_{Ar} = \frac{(2/1000)^3\ (0.2804)\ (2699.37 - 0.28)\ (9.807)}{(4 \times 10^{-5})^2} = 37{,}118$$

(Difference due to rounding)

Made in United States
North Haven, CT
05 November 2026

21760257R00015